S T U D E N T GUIDE

EXPLORING THE UNKNOWN

WRITING AND SOLVING EQUATIONS

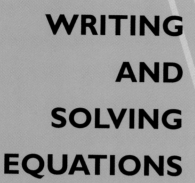

MathScape
SEEING AND THINKING
MATHEMATICALLY

How can
algebra
help you
analyze
and solve
problems?

EXPLORING THE
UNKNOWN

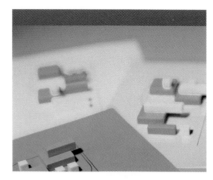

PHASE**ONE**
Working with Expressions

This phase introduces you to Lab Gear®, a tool for working with and understanding algebra. Lab Gear will help you simplify and evaluate algebraic expressions. You will also investigate ways to multiply expressions. At the end of the phase, you will use everything you have learned to analyze some number tricks.

PHASE**TWO**
Minuses and Parentheses

In algebra, there are three different uses of the minus sign. In this phase, you will explore these different uses and see how they are related. You will also work with expressions that involve parentheses. All of these ideas will come together at the conclusion of the phase when you describe a geometric pattern.

PHASE**THREE**
Solving Equations

In Phase Three you will use Lab Gear to model equations. Experimenting with Lab Gear will lead you to develop some algebra rules for solving equations. A final problem involving two different CD offers will give you a chance to apply everything you have learned.

PHASE ONE

23

In this phase you will learn to use Lab Gear, a tool for understanding and modeling algebra. By using Lab Gear, you will be able to simplify, evaluate, and multiply algebraic expressions.

Algebraic expressions are useful in describing everything from geometric patterns to scientific formulas. What are some ways in which you have used expressions in your previous mathematical work?

Working with Expressions

WHAT'S THE MATH?

Investigations in this section focus on:

ALGEBRA

- Understanding variables and using them to write expressions

- Evaluating expressions

- Simplifying expressions by combining like terms

- Multiplying two expressions

PATTERN SEEKING

- Writing expressions to describe number patterns

mathscape3.com/self_check_quiz

1 Gearing Up

In the same way that a protractor is a tool for working with and understanding angles, Lab Gear is a tool for working with and understanding algebra. You will experiment with Lab Gear blocks to see how they are named. Then you will use them to represent algebraic expressions.

Sort and Name Lab Gear Blocks

What are the names for the various Lab Gear blocks?

Your teacher will provide you with a set of Lab Gear blocks.

1 Sort the blocks in a way that makes sense to you.

2 The names of the blocks are based on what they represent.

  a. What do you think yellow blocks represent?

  b. What do you think blue blocks represent?

  c. What name could you give to each different type of block? The names of two blocks that represent *variables* are shown here to get you started.

 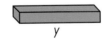
x y

Be ready to discuss your ideas with the class.

Measuring Lab Gear Blocks

You can use the corner piece to measure the length and width of Lab Gear blocks.

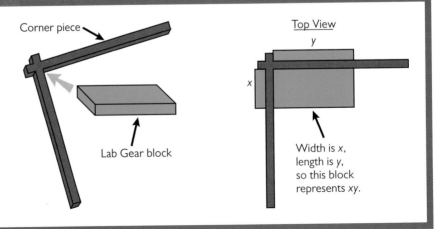

Corner piece

Lab Gear block

Top View
y

x

Width is x,
length is y,
so this block
represents xy.

Name Collections of Blocks

For this investigation, you will need a partner and a bag of Lab Gear.

1 Without looking, reach into the bag and pull out a collection of 6 blocks. Sketch a top view of the collection. Write an expression for this collection and then combine like terms to write the expression the short way. Repeat the process four more times, using collections of 7 blocks, 8 blocks, 9 blocks, and 10 blocks.

2 Write each of your expressions on a clean sheet of paper and trade this paper with another pair of students.

3 Put out blocks to match each of the expressions you are given. Then sketch the blocks next to the corresponding expression.

4 Trade back papers. Check to see if the other pair of students correctly sketched blocks to match each expression.

> **How can writing expressions for collections of blocks help you combine like terms?**

How to Write an Expression and Combine Like Terms

The value of this collection is written $x^2 + 5 + x + x^2 + x + x^2$

OR

We can combine like terms: $3x^2 + 2x + 5$

hot **words** | expression
like terms

Homework
page 210

2 What's in the Bag?

Blue Lab Gear blocks represent variables. What happens if you know the value of these variables? You will use Lab Gear to explore this question. Then you will be ready to use what you know about evaluating expressions, as well as some logical reasoning, to figure out which Lab Gear blocks are hidden in a bag.

Evaluate Expressions Using Lab Gear

How can you evaluate expressions when you know the value of variables?

Write an expression for each collection of blocks. Then use Lab Gear to help evaluate each expression for the given values of the variables. Keep a written record of your work.

1

a. $x = 1$ **b.** $x = 4$ **c.** $x = \dfrac{1}{2}$

2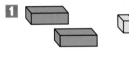

a. $x = 0$ **b.** $x = 6$ **c.** $x = 1$

3

a. $y = 1$ **b.** $y = 3$ **c.** $y = 5$

4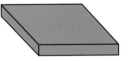

a. $x = 1, y = 2$ **b.** $x = 3, y = 1$ **c.** $x = \dfrac{1}{4}, y = 3$

Were you always able to replace the variable blocks with constant blocks? If not, how did you evaluate the expressions? Write a short summary of your process.

Play "What's in the Bag?"

Follow the directions below to play several rounds of "What's in the Bag?" Different pairs of students in your group should take turns being the bagkeepers. The other students try to guess which blocks are in the bag.

As you play, keep track of any strategies that you use to guess what is in the bag.

What strategies can you use to guess which blocks are in the bag?

Rules for "What's in the Bag?"

- Two players, the bagkeepers, secretly place no more than four Lab Gear blocks in a bag.

- Other players take turns naming values for the variables x and/or y.

- The bagkeepers carefully evaluate the contents of the bag using these values. Then they tell the final result to the other players.

- The players continue to name values for the variables until they are able to guess correctly which blocks are in the bag.

Write About Strategies

Write a paragraph describing any strategies you used to play "What's in the Bag?" Include a discussion of how you and your partners figured out the contents of the bag in one round.

hot **words** | expression

HW**omework**

page 211

3 Cornered!

MULTIPLYING EXPRESSIONS

You know how to multiply numbers, but can you multiply algebraic expressions? As you will see, the corner piece is very helpful when you need to multiply two expressions. In fact, knowing how to multiply with the corner piece will help you identify some common errors that algebra students make.

Multiply Two Expressions

How can you use the corner piece to multiply two algebraic expressions?

Work with a partner for the following.

1 What multiplication problem is shown here? Write this algebraically.

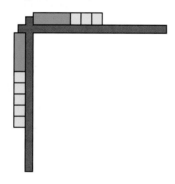

2 Use Lab Gear blocks to make a rectangle that fits inside the corner piece. The length and width of your rectangle should exactly match the length and width shown by the blocks on the outside of the corner piece. Sketch your rectangle.

3 What collection of blocks did you use to make your rectangle? What expression do they represent?

4 Write an equation that relates the original multiplication problem and the final product.

Identify Common Errors

These equations show some multiplication problems that were done by a student. There are at least one or two common errors here. Can you find them?

How can Lab Gear help you identify some common errors in multiplication?

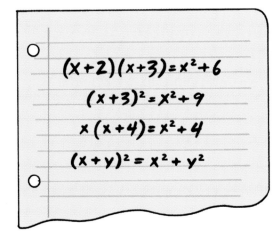

$$(x+2)(x+3) = x^2 + 6$$
$$(x+3)^2 = x^2 + 9$$
$$x(x+4) = x^2 + 4$$
$$(x+y)^2 = x^2 + y^2$$

Use Lab Gear to do each multiplication problem. Decide whether each equation is true or false. If an equation is false, rewrite the equation correctly.

Keep a written record of your work. Include a sketch of the Lab Gear blocks for each problem.

hot **words** | multiplication
rectangle

Homework
page 212

4 Lab Problem No. 1

This is your chance to be a number magician. You will start by exploring some number tricks. Then you will see how to use what you know about Lab Gear and algebra to help explain why the tricks work!

Explore a Number Trick

What do you notice about the results of a number trick?

Choose one of the number tricks below. Experiment with different sets of numbers until you see the trick.

Keep a written record of your work. Be ready to describe your trick to a partner.

Number Tricks

1. Inner and Outer Products

Choose any four consecutive whole numbers. Find the product of the inner numbers and the product of the outer numbers. What do you notice? Does this always work?

Inner product is 6 • 7

5 6 7 8

Outer product is 5 • 8

2. Calendar Squares

Choose any four numbers that form a square on a calendar. Find the product of the two diagonals. What do you notice? Does this always work?

Sun	Mon	Tue	Wed	Thu	Fri	Sat
	1	2	3	4	5	6
7	8	9	10	11	12	13
14	15	16	17	18	19	20
21	22	23	24	25	26	27
28	29	30				

The diagonal products are 3 • 11 and 10 • 4

Investigate Why It Works

Use algebra to show that your number trick always works.

1 Think about the four numbers in your number trick. Suppose the smallest number is represented by *x*. What are the values of the other numbers?

2 How could you write the products in your number trick using algebra?

3 Use Lab Gear, or other methods, to help you do the multiplication.

4 How do your results show that the number trick always works?

How can you use algebra to show that your number trick always works?

Write About the Results

Write a summary of your results. Include the following:

- a description of the number trick

- a summary of how you used algebra to show that the trick always works

- a summary of how you used Lab Gear or other techniques along the way

hot **words** | product

Homework
page 213

PHASE TWO

Minus signs are used in several different ways in mathematics. In this phase, you will use Lab Gear to help explore these different meanings of minus signs. You will also see how expressions can be written with and without the use of parentheses.

At the end of this phase, you will apply everything you have learned to describe a geometric pattern.

Minuses and Parentheses

WHAT'S THE MATH?

Investigations in this section focus on:

NUMBER

- Working with models for signed-number operations

ALGEBRA

- Simplifying polynomials

- Writing expressions with and without parentheses

- Using the distributive property to simplify expressions

PATTERN SEEKING

- Describing and analyzing a geometric pattern

MathScape Online
mathscape3.com/self_check_quiz

5 Extending the Lab Gear Model

Do you think you can use Lab Gear to show the expression −5 − 4? So far, you have only used Lab Gear to represent positive quantities. Now you will explore two ways to show minuses and negatives.

Use the Workmat to Add and Subtract

How can you model addition and subtraction of integers?

Work with a partner for each of the following. Record all of your steps on the Workmat Recording Sheet.

1 Use Lab Gear to model each addition problem.

　　a. −6 + 4　　　　b. −8 + 10　　　　c. −6 + (−3)

2 Create and solve addition and subtraction problems of your own. Be ready to share them with the class.

3 Use Lab Gear to model each subtraction problem.

　　a. −7 − (−3)　　　　b. 5 − 9　　　　c. −4 − (−9)

Adding and Subtracting with Lab Gear

To add −5 + 2...

Put down the first number	Put down the second number	Cancel what you can	Count what remains
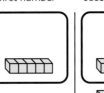			−5 + 2 = −3

To subtract −8 − (−2)...

Put down the first number	Take off the second number	Count what remains
−8	− (−2)	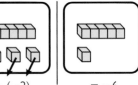 = −6

Show the Number 3 in Different Ways

Use Lab Gear to show (or "model") the number 3 in as many ways as possible.

- Use the minus area in some of your models.

- Use upstairs blocks in some of your models.

- Keep a record of your work using the Workmat Recording Sheet. Also, write an expression that corresponds to each model.

How many different ways can you show a number using Lab Gear?

Showing Numbers in Different Ways

Another way to "show minus" with Lab Gear is by putting blocks "upstairs" (on top of other blocks). You can use the upstairs blocks and the minus area to show the same number in a variety of ways. Here are three ways to show −4.

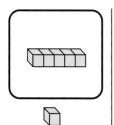

Negative 5 plus 1
−5 + 1 = −4

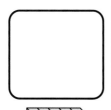

1 minus 5
1 − 5 = −4

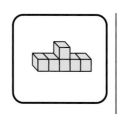

The opposite of 5 − 1
−(5 − 1) = −4

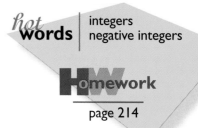

hot **words** | integers
negative integers

Homework
page 214

6 Making Long Expressions Shorter

SIMPLIFYING POLYNOMIALS

How can you write a long expression in a simpler way? You have already seen how an expression like $x + x + x + x$ can be written as $4x$. Now you will use Lab Gear to help simplify special types of expressions called polynomials.

Create and Simplify Polynomials

What rules can you find for simplifying polynomials?

Work with a partner for the following.

1 Choose as many blocks as you like of each type shown below. Place some blocks inside the minus area of your workmat and some outside the minus area. Do not stack the blocks.

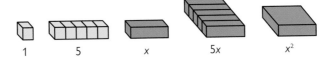

1   5 x $5x$ x^2

2 Sketch your collection on the Workmat Recording Sheet. Then write a polynomial for the collection of blocks.

3 Simplify the polynomial by canceling and clustering matching blocks.

4 Sketch the simplified collection on the Workmat Recording Sheet. Then write an expression for the simplified polynomial.

Repeat at least five times. Compare your original polynomials to the simplified ones. Write down any algebra rules that you could use in the future.

Simplify More Complex Polynomials

Do the following steps for each collection of blocks shown here.

What additional rules can you find for simplifying polynomials?

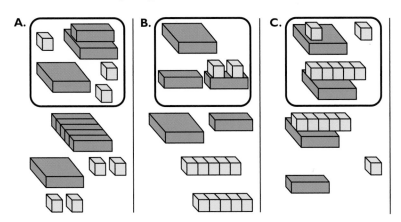

1 Put out blocks to match each collection. Write the polynomial for the collection.

2 Simplify the collection of blocks. Sketch the result on a Workmat Recording Sheet.

3 Write the simplified polynomial.

Write down any algebra rules that you could use in the future to simplify polynomials.

Polynomials
A product of numbers and variables is called a **term.** Examples:
$3x \qquad -2xy \qquad \frac{3}{8}y^2 \qquad 17 \qquad 5x^5$
A **polynomial** is an expression made up of the sum or difference of terms. Examples:
$3x + 5 \qquad 14x^2 + 5x - 2xy \qquad -y^2 - 6y + \frac{1}{2}$

hot **words** | term
expression

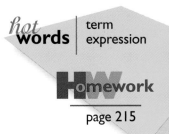

Homework

page 215

7 Grouping and Ungrouping

Is 3(y + 4) the same as 3y + 4? Parentheses can be used to group numbers and expressions, but you have to be careful if you want to remove the parentheses. You will be exploring some ways to write expressions with and without parentheses.

How can you write an expression without parentheses?

Decide Whether Equations Are True or False

The student who wrote these equations wasn't sure whether the two sides of each equation were really equal.

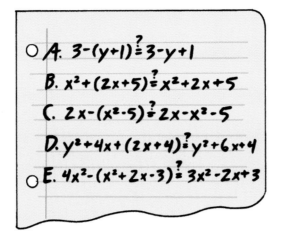

A. $3-(y+1) \overset{?}{=} 3-y+1$

B. $x^2+(2x+5) \overset{?}{=} x^2+2x+5$

C. $2x-(x^2-5) \overset{?}{=} 2x-x^2-5$

D. $y^2+4x+(2x+4) \overset{?}{=} y^2+6x+4$

E. $4x^2-(x^2+2x-3) \overset{?}{=} 3x^2-2x+3$

1. For each equation, show both sides using Lab Gear.

2. Record your setup on the Workmat Recording Sheet.

3. Decide if the two sides of the equation are equal.

4. If an equation is false, rewrite the right side so that it is true.

When can you remove parentheses without changing the value of an expression? What rules can you state for writing expressions without parentheses?

Explore the Distributive Property

Work with a partner to write each of the following expressions without using parentheses.

How can you use the distributive property to write expressions without parentheses?

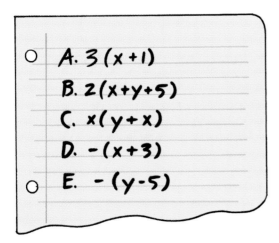

○ A. $3(x + 1)$

B. $2(x + y + 5)$

C. $x(y + x)$

D. $-(x + 3)$

○ E. $-(y - 5)$

1 Use Lab Gear to show each expression. For some expressions, you may need to use the corner piece; for others, you may need to use the minus area of the workmat.

2 Use Lab Gear to help you find a way to write each expression without using parentheses.

The expressions that you worked with all involve examples of the **distributive property.** How would you complete the following? "The distributive property says that $a(b + c) = $ _____."

Write About Parentheses

Write a summary that you can refer to later if you need to remember how to work with parentheses. Include the following:

- examples of expressions that are written with and without parentheses

- a description of the distributive property

- any other rules that are useful in working with parentheses

hot **words** | distributive property

page 216

8 Lab Problem No. 2

DESCRIBING A
GEOMETRIC
PATTERN

What does algebra have to do with geometric patterns?

Here is a chance to find out. First you will describe a pattern of blue and white tiles. Then you will use everything you have learned about writing and simplifying expressions to check your description.

How can you use algebra to describe a geometric pattern?

Describe a Patio Pattern

Consider the pattern below. Suppose the length of the patio is *x*. Write an algebraic expression for each of the following:

1 the total number of tiles needed

2 the number of blue tiles needed

3 the number of white tiles needed

A Pattern of Tiles

Small white tiles are placed together to make a large square patio. Then the white tiles are surrounded by a border of blue tiles.

Patio of
length 1

Patio of
length 2

Patio of
length 3

Check the Expressions

If the expressions you wrote to describe the patio pattern are correct, then when you take the expression for the total number of tiles and subtract the expression for the number of blue tiles, the result should equal the expression for the number of white tiles.

As you work, be sure to keep a written record of your process. Use the Workmat Recording Sheet to record any Lab Gear work.

1 Write an expression for the total number of tiles minus the number of blue tiles.

2 Simplify the expression.

   a. You may want to start by using the corner piece to multiply some expressions.

   b. You may want to use the workmat to help represent and simplify the resulting expression.

3 What is the final resulting expression? Is it equal to your expression for the number of white tiles?

How can you check the expressions you wrote to describe the tile pattern?

Write About the Process

Write a summary of your work that includes the following:

- a description of how you used algebra to show that your patio tile expressions were correct

- a summary of how you simplified expressions and a description of how you used the distributive property

- a summary of how you used Lab Gear or other techniques along the way

*hot*words | square expression

Homework
page 217

PHASE THREE

At the beginning of this phase, you will explore ways to add and subtract polynomials. These ideas will be helpful in the rest of the phase when you develop rules for solving equations. The phase ends with an opportunity to compare two CD offers. This will give you a chance to use everything you have learned about expressions and solving equations.

Solving Equations

WHAT'S THE MATH?

Investigations in this section focus on:

ALGEBRA

- Adding and subtracting polynomials

- Solving linear equations

- Writing expressions and equations to help solve problems

PATTERN SEEKING

- Using tables and patterns to help analyze a real-world problem

MathScape Online
mathscape3.com/self_check_quiz

 **Polynomial Arithmetic**

You have already seen how to use Lab Gear to add and subtract integers and simple expressions. Now you will use it to help add and subtract polynomials.

Model a Subtraction Problem

How is subtracting a polynomial related to addition?

Use Lab Gear blocks to model the following subtraction problem: $(x^2 - 4x + 4) - (5x - 3)$. Try to find two different ways to show this with Lab Gear.

When you have found the result, complete this sentence: "To subtract $(5x - 3)$, we ended up adding ____."

Adding and Subtracting Polynomials

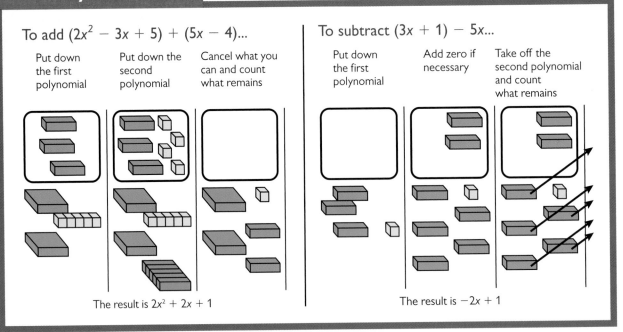

To add $(2x^2 - 3x + 5) + (5x - 4)$...

Put down the first polynomial | Put down the second polynomial | Cancel what you can and count what remains

The result is $2x^2 + 2x + 1$

To subtract $(3x + 1) - 5x$...

Put down the first polynomial | Add zero if necessary | Take off the second polynomial and count what remains

The result is $-2x + 1$

Add and Subtract Polynomials

Use Lab Gear to help you do each of the following addition or subtraction problems. Keep a written record of your work.

One of the six expressions cannot be simplified. Which expression is it and why?

A. $(4x+2)+(3x+1)$

B. $2y-7y$

C. $3xy-(-2xy)$

D. $2xy-2x$

E. $(2y^2+5y-2)-(y+8)$

F. $(x^2-6x+4)+(2x-3)$

How can you use polynomial addition and subtraction to simplify expressions?

Write Polynomial Arithmetic Rules

Write a summary of any algebra rules you could use in the future to add and subtract polynomials. Include some specific examples in your summary.

hot **words** | expression
subtraction

Homework

page 218

10 Simplify and Solve

SOLVING LINEAR
EQUATIONS

Now you are ready to use everything you have learned to help you solve equations. You will start by setting up an equation with Lab Gear and simplifying both sides to find a solution. Then you will begin to develop your own rules for solving equations.

Set Up and Solve an Equation

How can you model an equation to find a solution?

Work with a partner for this investigation. Record each step of your process on a Workmat/Equation Recording Sheet.

1 Use Lab Gear to show this equation on a workmat.

$$3x - 2 - (x - 2) = 2x + 6 - (2x + 2)$$

2 Simplify each side of the equation. Write the resulting equation.

3 What value of x makes the two sides equal? How do you know?

4 Check that this value of x is a solution by substituting it into the original equation.

Solve Some Equations

Work with classmates to solve each of the following equations.

How can simplifying each side of an equation help you find a solution?

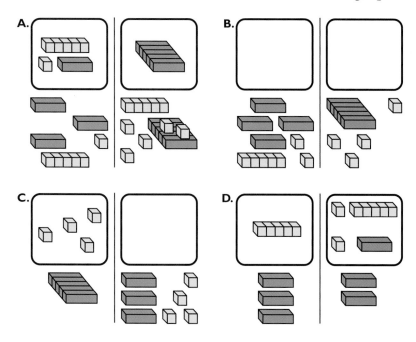

1 Write the starting equation on your Workmat/Equation Recording Sheet.

2 Use Lab Gear to solve the equation. Use the Workmat/ Equation Recording Sheet to keep track of some of the equations and Lab Gear positions along the way.

3 Check your solution in the original equation.

Did you and your classmates find any rules or shortcuts for working with Lab Gear? Be ready to discuss these with the class.

hot **words** | equation solution

Homework page 219

11 Staying Balanced

How do Lab Gear rules correspond to algebra rules?

You will see how thinking of an equation as a balanced scale can help you develop new Lab Gear methods. Then you will translate these methods into algebra rules and solve some equations without using Lab Gear.

Solve Equations with Lab Gear

What rules can you use to keep an equation balanced and find its solution?

To solve the following equations, it may be helpful to think of each equation as a balanced scale. You can add or remove the same Lab Gear blocks from both sides to keep the equation balanced.

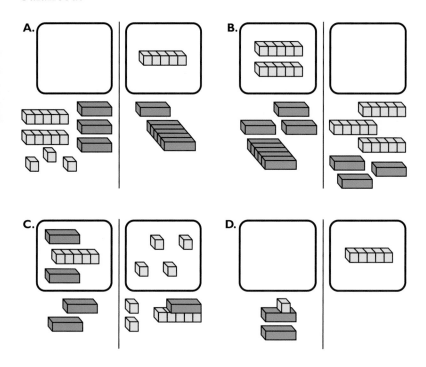

1 Write the starting equation on your Workmat/Equation Recording Sheet.

2 Use Lab Gear to solve the equation. Use the Workmat/Equation Recording Sheet to keep track of some of the equations and Lab Gear positions along the way.

3 Check your solution in the original equation.

Did you and your classmates find any rules or shortcuts for working with Lab Gear? Be ready to discuss these with the class.

Solve Equations Without Lab Gear

How can you solve equations using only paper and pencil?

Work with a partner to solve the following equations:

1 Solve each equation without using Lab Gear. Keep a careful record of each step of your process.

2 Check your solutions by substituting them into the original equations.

$$\text{A. } 16 = 3x - 2$$
$$\text{B. } 4x + 1 = 6x - 9$$
$$\text{C. } -y + 5 = 6y + 12$$
$$\text{D. } 3x - (2x + 1) = 3x + (5 - 4x)$$
$$\text{E. } 2(y + 4) = 2y + y + y + 14$$

What algebra rules did you use to help solve the equations? Be ready to share your methods with the class.

Write About the Process

Write a summary of everything you know about solving equations. Include the following:

- algebraic rules or methods you could use in the future to help solve equations

- specific examples of equations and how you can solve them

- a discussion of how you can check a solution

hot **words** | equation
solution

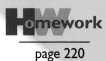

Homework

page 220

12 Lab Problem No. 3

Have you ever seen an advertisement offering 15 CDs for one dollar? Sometimes these deals are not quite what they seem. You will have a chance to compare two different CD offers to find out which is the better deal. Your equation-solving skills will come in handy!

Write Expressions to Describe the Offers

What expressions can you write to describe each CD offer?

Work with a partner to describe and compare the two magazine advertisements shown below. Let x be the total number of CDs you buy.

1 Write an expression that describes your total cost if you get the CDs from MusicMania.

2 Write an expression that describes your total cost if you get the CDs from DiscZone.

MusicMania
- Order any number of CD's.
- We'll give you the first 2 of them free!
- Pay just $9 for each of the remaining CDs.

DiscZone
- CDs are just $7 each!
- One-time membership fee of $4.

Find the "Break-Even" Point

How many CDs would you have to buy to make the cost the same under both plans? This is the "break-even" point. You can find the break-even point by setting your two expressions equal to each other.

1 What equation do you get by setting the two expressions equal?

2 Solve the equation. Keep a step-by-step record of your work.

3 Check your solution.

What does your result tell you about the two CD offers?

How can solving an equation help you compare the two CD offers?

Compare the Two Offers

Write a comparison of the two CD offers. Be sure to answer the following questions and include a discussion of how you arrived at your conclusions.

- When is MusicMania a better deal? When is DiscZone a better deal?

- How many CDs do you need to buy to make the cost the same under both plans?

- Describe how you set up and solved an equation. Discuss the methods you used to solve the equation.

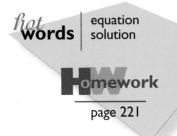

hot **words** | equation
solution

HW **omework**
page 221

Gearing Up

Applying Skills

Write an expression for each collection of Lab Gear blocks. Then combine like terms to write the expression the short way.

1.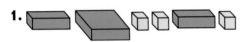

2.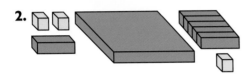

Simplify each algebraic expression by combining like terms.

3. $x + y + x$

4. $y + y + x + 3$

5. $y^2 + y + y^2 + y^2$

6. $x^2 + y^2 + 1 + x^2$

7. $x + x^2 + x + x$

8. $x^2 + y^2 + y^2 + x^2$

9. $5 + x + y + 6$

10. $2y + xy + x + 3y$

Tell whether the two expressions in each pair are equivalent.

11. $x + x + y + x$ and $3x + y$

12. $x^2 + y + y^2$ and $x^2 + 2y^2$

Extending Concepts

Simplify each expression by combining like terms. Explain how you figured out each answer.

13. $7.3x^2 + x + 2.4x^2 + 4.3x + 1.1x^2$

14. $\frac{1}{2}x^2 + xy + \frac{1}{4}x^2$

15. Marta pulled out a collection of Lab Gear blocks from a bag. She wrote an expression for the collection. Then she combined like terms and wrote the expression the short way. The short expression was $7y + 6$.

What could the collection of blocks have been? List as many different possibilities as you can. Assume that the bag contained lots of blocks of each type.

Writing

16. Answer the letter to Dr. Math.

Dear Dr. Math:

I want to simplify the expression $3y^2 + xy + y + 4y^2$ by combining like terms. How do I know which terms I can combine? Once I know which terms I can combine, how do I combine them? Can you explain the method so that I can figure it out myself in the future?

Pete

What's in the Bag?

Applying Skills

Evaluate each expression for the given values of the variables.

1. Evaluate $3x + 1$ for:

   a. $x = 1$ **b.** $x = 3$ **c.** $x = 5$

2. Evaluate $y^2 + 2$ for:

   a. $y = 0$ **b.** $y = 1$ **c.** $y = 4$

3. Evaluate $x^2 + 4x$ for:

   a. $x = 1$ **b.** $x = 2$ **c.** $x = 6$

4. Evaluate $4y - 6$ for:

   a. $y = 0$ **b.** $y = 4$ **c.** $y = \frac{1}{2}$

5. Evaluate $x^2 + x + 3$ for:

   a. $x = 0$ **b.** $x = 1$ **c.** $x = 2$

6. Write an expression for the collection of Lab Gear blocks below. Then evaluate the expression for the given values of the variables.

   a. $x = 2$ **b.** $x = 5$ **c.** $x = 0.2$

Extending Concepts

7. In a game of "What's in the Bag?" José puts Lab Gear blocks in a bag. Julia names values for the variable, and José evaluates the contents of the bag. Figure out which blocks are in the bag. (Hint: There are 3 blocks in the bag and only 1, 5, x, and $5x$ blocks are allowed.)

Values	Result
$x = 0$	1
$x = 1$	7
$x = 2$	13
$x = 3$	19

Making Connections

8. The formula for the volume of a cylinder is $V = \pi r^2 h$, where r represents the radius and h the height. What is the volume of a cylinder if its radius is 3 cm and its height is 1.4 cm? How did you figure it out? How is this similar to evaluating expressions?

Cornered!

Applying Skills

1. a. What is the multiplication problem shown below? Write this algebraically.

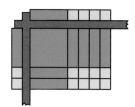

b. Write an expression for the collection of blocks used to make the rectangle inside the corner piece.

c. Write an equation that relates the multiplication problem and the final product.

For each multiplication problem, make a sketch of the Lab Gear setup. Then write an equation that shows the result of the multiplication.

2. $x(x + 3)$

3. $(x + 2)^2$

4. $(x + 1)(x + 4)$

5. $(x + 2)(x + 5)$

Extending Concepts

6. Suppose that you have the Lab Gear blocks shown here.

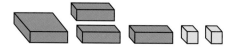

a. Write an expression to represent the collection of blocks.

b. Find a way of arranging these blocks to form a rectangle. Make a sketch of the arrangement.

c. What multiplication problem does this arrangement show? Write an equation that relates the multiplication problem and the expression in item **6a.**

Making Connections

7. The sum of the first x even numbers is equal to $x(x + 1)$.

a. How can you write this expression without using parentheses?

b. What is the sum of the first 50 even numbers? How did you find it?

Lab Problem No. 1

Applying Skills

Simplify each algebraic expression by combining like terms.

1. $x + y + x + x$ **2.** $y^2 + y + y^2 + xy$

Evaluate each expression for the given values of the variables.

3. Evaluate $y^2 + 5$ for:

   a. $y = 1$ **b.** $y = 3$ **c.** $y = 7$

4. Evaluate $x^2 + 3x + 2$ for:

   a. $x = 0$ **b.** $x = 2$ **c.** $x = 6$

Make a sketch of Lab Gear to do each multiplication problem.

5. $(x + 1)^2$ **6.** $(x + 3)(x + 4)$

Extending Concepts

7. Leanne and Jody each gave an argument to support this statement: "If you pick any four consecutive even numbers, the inner and outer products will differ by 8." Which argument do you find more convincing? Why?

Inner product is 4 • 6

2 4 6 8

Outer product is 2 • 8

Leanne: "I tried lots of different sets of four consecutive even numbers, and the outer product and the inner product always differed by 8."

Jody: "I called the numbers x, $x + 2$, $x + 4$, $x + 6$. The two products are $x(x + 6)$ and $(x + 2)(x + 4)$. These can be written as $x^2 + 6x$ and $x^2 + 6x + 8$. This shows that whatever the numbers are, the products differ by 8."

Making Connections

8. In the old Chinese calendar, each month was made up of three 10-day periods similar to our week.

1	2	3	4	5	6	7	8	9	10
11	12	13	14	15	16	17	18	19	20
21	22	23	24	25	26	27	28	29	30

The figure shows a calendar with 10 days in each row. Choose any four numbers that form a square. Find the products of the two diagonals. Repeat this for other blocks of four numbers. Describe what you notice.

Extending the Lab Gear Model

Applying Skills

What number is shown by each collection of Lab Gear blocks?

1.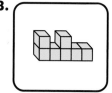

2.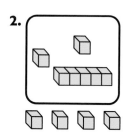

3.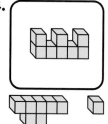

4.

Solve each addition or subtraction problem.

5. $-8 + 3$

6. $5 - 9$

7. $-8 + (-6)$

8. $-4 + 11$

9. $-3 - (-9)$

10. $-(10 - 3)$

11. $-(-7 + 1)$

12. $-(-5 - 7)$

Extending Concepts

13. a. Sketch three different Lab Gear models for the number 4. In one use the minus area, in the second model use upstairs blocks, and in the third use both the minus area and upstairs blocks.

b. What number could you subtract from 4 to end up with 7? How did you figure this out?

Making Connections

14. The highest temperature ever recorded in the United States was 134°, recorded in Death Valley, California. The lowest temperature was $-80°$, recorded in Alaska. Write and solve a subtraction problem to find the difference between the highest and lowest recorded temperatures.

15. Mauna Kea is a volcano in Hawaii. It is the highest island mountain in the world. Its peak is about 13,800 ft above sea level. Its base is about 18,200 ft below sea level. Write and solve a subtraction problem to find the approximate height of Mauna Kea from its base to its peak.

Making Long Expressions Shorter

Applying Skills

Tell whether each expression below is a polynomial. Explain your answers.

1. $2x^2 + 8$

2. $\sqrt{x} - 2$

3. $7x - 1 + x^3$

4. $\dfrac{1}{x}$

Simplify each polynomial.

5. $5x^2 + 7x - 2x$

6. $4x^2 + 3x + 1 - 2x^2$

7. $8y + 5 - y + 3y^2$

8. $7x^2 + x + 4 - 6x^2 - 3$

9. Write a polynomial for this collection of Lab Gear blocks. Then write the simplified polynomial.

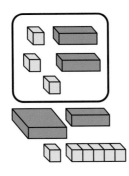

Extending Concepts

10. a. Write a polynomial for this collection of Lab Gear blocks.

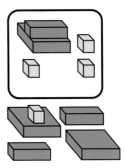

b. Sketch the simplified collection of blocks.

c. Write the simplified polynomial.

Making Connections

11. The number of diagonals in a regular polygon with x sides is $0.5x^2 - 1.5x$.

a. Is this expression a polynomial? How can you tell?

b. How many diagonals does a regular polygon have if it has 8 sides? 12 sides? How did you figure out your answers?

Grouping and Ungrouping

Applying Skills

Tell whether each equation is true or false. If it is false, rewrite the right side so that it is true.

1. $8 + (x + 2) \overset{?}{=} 8 + x + 2$

2. $6 + (y + 7) \overset{?}{=} y + 13$

3. $x^2 - (x + 6) \overset{?}{=} x^2 - x + 6$

4. $2y^2 - (y + 3) \overset{?}{=} 2y^2 - y - 3$

5. $-(x + 6) \overset{?}{=} -x - 6$

6. $-(y - 1) \overset{?}{=} -y - 1$

Write each expression without using parentheses.

7. $1 + (x + 7)$ **8.** $y^2 + (y - 3)$

9. $x^2 - (x + 2)$ **10.** $5y^2 - (y - 2)$

11. $6x + (5x - 1)$ **12.** $9y - (2y + 3)$

13. $-(x + 9)$ **14.** $-(y - 6)$

Extending Concepts

15. Two students, Vinh and Danielle, explained how they would write the expression $5x^3 - (2x^2 + 3x - 1)$ without parentheses.

Vinh: "Remove the parentheses and make all the terms that were inside the parentheses negative."

Danielle: "Remove the parentheses and switch the signs of all the terms that were inside the parentheses."

Which method do you think is correct? Why? What is the result when the expression is written without parentheses?

Making Connections

16. When a manufacturer sells x units of a product, its income is $12x$. Its expenses for the x units are $x^2 - 8x + 40$. The profit can be found by subtracting the expenses from the income as follows: $12x - (x^2 - 8x + 40)$.

a. Why does it make sense that profit is found by subtracting expenses from income?

b. Write the expression for profit without parentheses and simplify.

Lab Problem No. 2

Applying Skills

Simplify each polynomial.

1. $6x^2 + x - 2x^2 + 3$

2. $4x^2 + 3x + 1 - 8x$

3. $3x^2 - 2x + 5 - x$

4. $5x^2 + x - 4 - 3x^2 - 5$

5. $-2x^2 + 4x + 1 + x^2 - x$

6. $2x - 3 - 6x + x^2$

Tell whether each equation is true or false. If it is false, rewrite the right side correctly.

7. $4y + (5y + 2) \overset{?}{=} 9y + 2$

8. $3x^2 - (x + 6) \overset{?}{=} 3x^2 - x - 6$

9. $6y^2 - (y - 4) \overset{?}{=} 6y^2 - y - 4$

10. $6y^2 - (2y^2 + y) \overset{?}{=} 4y^2 + y$

Write each expression without using parentheses.

11. $3x + (9x + 1)$

12. $x^2 - (x + 8)$

13. $5(x + 3)$

14. $-(y - 7)$

Extending Concepts

15. A rectangle is cut from the corner of a square piece of paper measuring 10 in. by 10 in. The length of the rectangle is one inch more than its width.

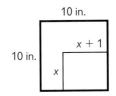

a. What is the area of the original square?

b. Use x to represent the width of the rectangle. Write an expression for the area of the rectangle that is removed.

c. Write the expression in item **15b** without parentheses. What rule did you use?

d. Write an expression for the area of the paper after the rectangle has been removed. (Hint: Subtract your answer in item **15c** from your answer in item **15a**). Give your answer both with and without parentheses.

Writing

16. Write a summary of the rules you have learned for removing parentheses from expressions. Be sure to describe how to remove parentheses that are used with subtraction and those that are used with multiplication. Give an example of how to use each rule.

Polynomial Arithmetic

Applying Skills

Simplify each addition or subtraction problem.

1. $(2x + 3) + (6x - 2)$

2. $(5y + 3) - (4y + 1)$

3. $2x^2 - 3x^2$

4. $2x + 3 - (x + 5)$

5. $2x - 1 + (-x + 3)$

6. $3y^2 + 1 - (y^2 - 2)$

Complete each sentence.

7. Subtracting $(2x + 1)$ is equivalent to adding ____.

8. Subtracting $(4y - 2)$ is equivalent to adding ____.

Writing

10. Answer the letter to Dr. Math.

Dear Dr Math:

I was trying to simplify this subtraction problem: $(-4x - 3) - (2x - 1)$.
I figured that I could change it to an addition problem like this: $(4x + 3) + (2x - 1)$, which comes out to $6x + 2$.
My friend got $-6x - 2$ for the answer. Where do you think she went wrong?

Polly Nomial

Extending Concepts

9. What addition problem is modeled by the sequence of Lab Gear moves shown here? Make a sketch of the Lab Gear that will remain after you cancel what you can. What is the result?

Step 1: Put down the first polynomial

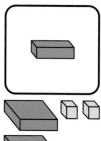

Step 2: Put down the second polynomial

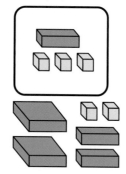

Simplify and Solve

Applying Skills

Determine whether the given value of x is a solution to the equation by substituting the value into the equation. Show your work.

1. $3x = 15$; $x = 5$

2. $2x - 6 = 20$; $x = 7$

3. $3x + 3 = 8 - (x + 1)$; $x = 1$

4. $4x + 5 - (x + 3) = 8 - (2x - 1)$; $x = 2$

5. a. Write the starting equation for the Lab Gear setup shown below.

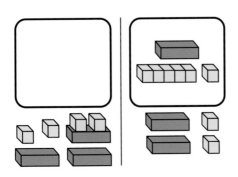

b. Find the value of x that makes the two sides of the equation equal.

c. Check your solution in the original equation.

Solve each equation. Check your solution in the original equation.

6. $2x = 14$

7. $y + 9 = 12$

8. $6x = -24$

9. $2x + 3 - (x + 3) = 7 - 5$

Extending Concepts

10. a. Write the starting equation for this Lab Gear setup.

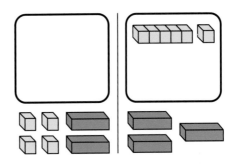

b. Solve the equation and check your solution in the original equation. How did you solve this problem?

Making Connections

11. The speed of light is about 11 million miles per minute. The distance of the earth from the sun is about 93 million miles. Let x represent the time (in minutes) that it takes for light from the sun to reach the earth. Then x satisfies the equation $11x = 93$.

a. Why does this equation make sense?

b. What value of x makes both sides approximately equal? To the nearest minute, how long does it take light from the sun to reach the earth?

Staying Balanced

1. a. Write the starting equation for the Lab Gear setup shown below.

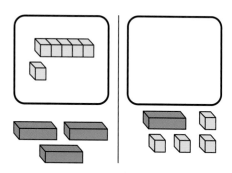

b. Solve the equation.

c. Check your solution in the original equation.

Solve each equation. Keep a record of each step of the process. Check your solution in the original equation.

2. $y - 4 = 8$

3. $18 = 4x + 2$

4. $3y - 4 = 20$

5. $6x - 2 = 16$

6. $5y - 3 = 2y + 9$

7. $x - 3 = 2x + 4$

8. $-y + 3 = 2y + 6$

9. $-2x + 10 = 2x + 2$

10. $4x - (x + 1) = x - (x - 5)$

11. $2(y + 3) = 16$

12. a. Sketch a sequence of Lab Gear setups to solve the equation $2x - (x + 4) = 4 - (x + 2)$.

b. Solve the equation using algebra rules. Keep a record of each step of the process.

c. Describe each of the Lab Gear rules that you used in item **12a**. Give the corresponding algebra rule from item **12b**.

13. If a ball is thrown straight up at a speed of 64 feet per second, its speed (in ft/sec) at x seconds after the toss will be $64 - 32x$. At what instant will the ball's speed be 0? (Hint: Solve $0 = 64 - 32x$.)

14. Celsius and Fahrenheit temperatures are related by the equation $F = 32 + 1.8C$. If the Fahrenheit temperature is 50°, what is the Celsius temperature? Show your work.

Lab Problem No. 3

Applying Skills

Solve each equation. Keep a record of each step. Check your solution.

1. $22 = 3x + 1$ **2.** $5x - 2 = 13$

3. $y + 3 = 4y - 9$ **4.** $6x + 3 = 4x + 1$

5. $-x + 14 = 2x + 2$ **6.** $2(y + 4) = 10$

For items **7–12**, compare schools **A** and **B**.

7. Which school would charge more if you took 10 lessons? 20 lessons?

> **School A**
> a $5 registration fee plus $10 per lesson
>
> **School B**
> $11 per lesson, the first lesson is free.

8. Write an expression showing the total cost for x lessons at School A.

9. Write an expression showing the total cost for x lessons at School B.

10. What equation do you get by setting the two expressions equal?

11. Solve the equation. Keep a record of each step. Check your solution.

12. For how many lessons will the cost be the same at Schools A and B?

Extending Concepts

> **Club A**
> First book is free, each additional book costs $8.
> **Club B**
> $5 registration fee, each book costs $9.
> **Club C**
> First two books cost $10 each, remaining books $6 each.

13. If you buy 10 books, which book club will charge the least? the most?

14. Write expressions for the total cost of x books at each book club.

15. How many books do you have to buy for the cost to be the same at Clubs A and C? Explain how you figured out your answer.

16. What equation do you get if you set the expressions for Clubs A and B equal? Solve the equation. How would you interpret the answer? Is there any number of books for which Club B charges less than Club A? Explain your thinking.

Writing

17. Answer the letter to Dr. Math.

> Dear Dr. Math:
>
> My teacher is always going on about expressions and equations. What's the difference anyway? And what does she mean when she asks us to solve an equation? Are there any rules that I can use to help me solve an equation?
> X. Pression

Mc Graw Hill **Glencoe**

This unit of MathScape: Seeing and Thinking Mathematically was developed by the Seeing and Thinking Mathematically project (STM), based at Education Development Center, Inc. (EDC), a non-profit educational research and development organization in Newton, MA. The STM project was supported, in part, by the National Science Foundation Grant No. 9054677. Opinions expressed are those of the authors and not necessarily those of the Foundation.

CREDITS: Photography: Chris Conroy • © Shiniichi Eguchi/Photonica: pp. 179TL, 180.

Creator of the Lab Gear®: Henri Picciotto.

Send all inquiries to:
Glencoe/McGraw-Hill
8787 Orion Place
Columbus, OH 43240-4027

ISBN: 0-07-866828-X

4 5 6 7 8 9 10 058 06 05